AUTHOR: AARON DAVID HATCHER

SCIENTIFIC THEORIES

MAY 7, 2014 - Current

Theories   Page

1 - 150   1 - 150

1

SEED BEHAVIOR AND DISTRIBUTION

1.   I saw a seed fly into the air.

2.   My hypothesis is that it detaches from the base and then is picked up by the wind at ten

miles per hour.

3.   I plan to use a fan and a flower to see how much breeze it takes to cause the detachment.

4.   Air speeds showed that at five, it stayed, at ten it flew and at fifteen miles per hour it also

flew. My hypothesis was correct.

5.   The new hypothesis is that it will travel one mile before landing, which I will discover.

*X 5 more experiments

Distance In Miles = D

Time in Minutes = T

Speed in MPH = S

(D10 * T60 = S)

•

2

MISSING ELEMENT

1.   I noticed that there are only so many elements on the periodic table.

2.   My hypothesis is that there are other elements.

3.   I plan on funding a mission to other planets to find more elements.

4.   The data is still being retrieved, and is not yet retrieved. It is not yet time to accept the hypothesis or not though it is very possible.

5. Even more planets are to be explored, as to collect more data and prove that there are in fact more elements.

Elements = E

118E + 1E = 119E

•

3

CAR POLLUTION

1. I smelled and saw a dark fume in the air by a highway.

2. My hypothesis is that the cars are making the smell and fume.

3. What I intend on doing is collecting the odor and fume, and then studying whether or not it had come from a car.

4. Studies showed that it was in fact from a car, and that it is polluting the air.

5.   Plans on making studies on other vehicles, such as helicopters, will prove to also be contributing to air pollution.

Polluting Properties = PP

Non Polluting Properties = NPP

$$PP4 + NPP4 = 4$$

$$/\,4 = /\,4$$

$$PP + NPP = 0$$

•

4

ICE INTO NEW SUBSTANCE

1.   I noticed that $H2O$, starts as steam at high temperatures, then water and ice.

2.   It is my hypothesis that ice, when cooled even further, creates a new substance, black hole or tear in time.

3.   I took water and heated it, then cooled it. It became steam, water and then ice. Then I tried to cool it even further.

4.   My hypothesis was suggestive, because I could not cool it to a cold enough temperature and so my findings were not sufficiently conclusive.

5.   I plan to instead cool it to a temperature twice that of what I had originally intended, to reach further conclusions.

Space in inches = SII

Temperature in degrees = TID

Result = R

$$SII2 + TID(-1000) = R$$

•

5

TIME CAN GO IN REVERSE

1.   I noticed that time goes in a certain direction. I also noticed that it goes in all directions.

2.   My hypothesis is that if a ball can go forward and backward in outer space, then time could also go forward and backward.

3.   By moving a ball back and forth, I noticed that time did not also go back and forth. The ball was tested.

4.   It was proven that a ball can move in all directions, but not necessarily time. Though it has not yet been proven that time moves in all directions, more tests will no doubt prove that my hypothesis was correct.

5.   Further studies will be conducted, because the theory must be true, by my prediction. Chemical reactions, outer space and mass are going to be subjected. Tests will be conducted, and my theory will be proven, granting motion of time in was before not at all possible.

*By mass, and by that I mean the milky way, could be equally effected by the other direction, thereby putting everything in the milky way in a reverse motion by another galaxy's either natural effect or intentional triggering.

Distance in inches = D

Time in seconds = T

$$D-1 + D1 + T60 = 0$$
$$D0 + T60 = 0$$
$$/\,60 = /\,60$$
$$T = 0$$

$$0 = 0$$

$$0$$

$$\bullet$$

$$6$$

TWO SEPARATE OBJECTS CAN BE EFFECTED BY ONE ANOTHER SIMULTANEOUSLY

1. I noticed that a vehicle down the street moved at the same time another object up the street.

2. My hypothesis is that both were effected by time, space and each others mass.

3. By means of video recording, I caught both objects moving at the same time, suggesting that they were influenced by one another.

4. My hypothesis was only theorized and hypothesized. If everything in the universe is moving at the same time, then everything is effected simultaneously and so it is proven.

5. By planning, I hope to conduct studies to show that though there is space between moving objects, that one can effect the other by means of a fabric of space more accurately.

Back and forth motion.

*If all of the mass, all of it, was moving together, then everything together effects everything

else.

*It should be possible to move distant objects through means of oxygen gas, which divides.

Distance Between Objects = DBO

String Theory Supporting Two objects and One Distance as One Object = ST

DBO20 = ST

/ 20 = / 20

DBO = ST / 20

* 20 = * 20

-------------------------->

DBO20 = ST

/ 20 = / 20

DBO = ST / 20

* 20 = * 20

<--------------------------

•

HEAT CAN BECOME TO SUCH A DEGREE NEVER DOCUMENTED

1. I have seen cool objects become warmer, hotter and then extremely hot. The sun is one of the objects seen in theorized infinite space.

2. Hypothetically it is thought that matter can reach a higher degree of heat, that which has never been seen before.

3. By heating water, the water temperature could be regulated in it's lows and also highs.

4. By conducting my experiment, I noticed that I am unable to make it hot enough, and so proving that it was possible, was suggestive.

5. By using explosives or monitoring the sun's mass, I would like to further my research on making hot hotter than it has ever been documented as but conducting studies.

Degree of heat = D

Matter = M

Energy = E

$$D + \infty = -1M + 1E + \infty$$

•

8

LIGHT USED AS FUEL AND ACCELERATOR

1.   I noticed that light reflects, moving around the universe.

2.   The hypothesis that I suggest is that if the right compilation of mass could interact with light causing it to move. If enough light was used, then it could, through a design, accelerate and then pose as a function such as a new flying machine.

3.   By monitoring the light from the sun, as in solar power, the intensities could be regulated. Once it is graphed, then more could be synthesized, and then the synthesized could be fueled by by the light, one supporting the other.

4.   My hypothesis is proven by tests conducted, subtly.

5.   Light is accessible, and in abundance. What I would like to do is study the possible uses of light, whether it be like a fuel, weapon or defense.

Light In Quantity = L

Circulation = C

$$L100 + 1000\% = C10{,}000\%$$

•

9

FOOD NOT NECESSARILY A NUTRIENT

1.   I have noticed that humans and animals eat food that might contribute to their death.

2.   My hypothesis is to dissect food, animals, and see if there are any correlations. A person's death could be directly influenced by what it digests, and it has been proven before. By comparing anatomic structures and how they relate, evidence could be found.

3.   Enough testing has been done at this time. I have bought food, and studied it as best as I could.

4.   Conclusions could be obtained, and so poisons are all that is known at this time, as well as unhealthy foods. Future studies will improve health, and so my hypothesis will most likely be proven true.

5.   I'd like to test subjects by using harmless measures, to arrive at more conclusions.

Nutrition Intake = N

Total Nutrition Residing = T

$$N10 + N(-10) = T$$

•

10

PEOPLE EVIDENCE OF ALIENS DUE TO BODY

1.   I have seen human anatomy, animal anatomy and insect anatomy and noticed that all of them are similar.

2.   My hypothesis is that that of possibility. I have seen anatomy, and it is designed in such a way that suggests that similar structures could have formed under the right circumstance such as O and H_2O.

3.   By taking O and H_2O in an isolated environment and then accelerating growth, I plan on creating life that could be on another world.

4.   My studies and tests are somewhat conclusive, and further research is needed in this field

before there will be any more results.

5.   With chemicals and compounds from our solar system and galaxy, I would like to do

other studies, like with different or new chemicals.

●

Chemical One Unit = CO

Chemical Two Unit = CT

Chemical Four Unit = CF

$$CO + CT = CF$$

11

ATOMIC DEVELOPMENT

1.   The atomic structure could be synthesized into creating any matter desired.

2.   My hypothesis is that I will be able to synthesize organic matter, and experimental matter

which could be used in redesigning and growing never before redesigned or grown material.

3.   By taking an atom and experimenting with it by combining and adjusting it's structure, I will do so with certain equipment, standard or experimental.

4.   Not having the equipment, my studies have been put temporarily on hold.

5.   Further testing biologically and elementally will be what my aim is.

Atom One = AO

Atom Two = AT

New Substances = NS

AO + AT = NS

•

12

PLATE TECTONICS IS MOVING TO REFORM INTO ONE CONTINENT AGAIN

1.   I think that every continent will form to be one continent before the destruction of the planet.

2. Having seen pictures of the continents and read about them and how they had once been one continent, I think that it will once again occur and this is my hypothesis.

3. By observing the changes of the tectonic plates, I predict that their movement will take new directions, even newer ones that have been discovered.

4. The new data that I have gathered suggests that the plates are moving, but the direction is unknown.

5. By using more equipment, like devises such as satellites and aqua sensory mechanisms, I will conduct more tests.

$$Continent\ One = CO$$

$$Continent\ Two = CT$$

$$Continent\ Three = CTH$$

$$Continent\ Four = CF$$

$$One\ Continent = O$$

$$Time\ in\ Centuries = T$$

$$Distance\ in\ Miles = D$$

$$(CO + CT + CTH + CF)\,(D*T) = 1$$

$$COCTCTHCF\,(D600*T4) = 1$$

BIOLOGY COULD BE USED TO CREATE BRAND NEW SPECIES

1.   I have seen many different species of life, and all of it was thought to have been created under similar conditions.

    2.   My hypothesis is that more life can be created, like in a laboratory.

  3.   New life, with out causing suffering, is what I aim to do by using my equipment.

4.   By using the equipment that I have, I have been unable to create my own species, but in time I may be successful. The species I created was barely alive.

  5.   Three new species are thought to be created by myself, in the next century, and more tests.

Splicing Number One = DNAO

DNA Splicing Number Two = DNAT

New Species = NS

$$DNAO + DNAT = NS$$

•

14

BIOLOGY COULD BE USED TO HEAL AND CREATE LONGER LIFE

1.   Having seen skin heal, this is my observation.

2.   The observation suggests that it could be synthesized, and cause longer life, successful surgery, and world peace. This is my hypothesis.

3.   By studying skin, and then atomic structure, I plan on making an ointment that regenerates age and wounds to young and healthy skin.

4.   Arriving at the conclusion I was aiming for, I did create the ointment.

5.   I want to continue studies and hopefully create life that could last forever.

Chemical One = CO

Chemical Two = CT

Ointment = O

$$(CO) + (CT) = (O)$$

•

15

QUANTUM PHYSICS THEORY

1.   I have seen books about Quantum Mechanics and string.

2.   The book appeared to be made out of paper and ink, and so proving this is my hypothesis. Also it is my hypothesis that the book itself is made out of string, such as in super string theory.

3.   By taking strings of yarn, I noticed that it looked like DNA, and that it looked like fiber which was interesting. Using equipment, I tried to study the string.

4.   I was not able to see the string, even if I was looking right at it due to the size. Not being able to see it, lead me to realized that it was hard to prove though possible.

5.   There will be more tests, so to prove the strings exist, and that other things also do

supportively.

String Structure = SS

Atoms = A

Space = P

A + P = SS

•

16

ANYTHING IS POSSIBLE

1.   I have seen, heard and smelled millions of separate things all that I did not know were

possible until I witnessed them my self. Empiricism has been studied.

2.   My hypothesis is that if I have seen something for the first time, then there could be a

first time for everything.

3.   I'll gather data, and then analyze it using numbers and statistics.

4. I noticed that the numbers are usually different, but not always. By not arriving at consistently different numbers or the same, everything is not always possible.

5. Collecting numbers representing new occurrences and old ones will be continued.

$$\text{Possibilities} = P$$

$$P = \infty$$

•

17

RELIGION ONE

1. I have read out of books that Jesus walked on water and turned water into alcohol.

2. My hypothesis is that although the stories are not correct, the teachings are correct and there for were not always of example.

3. I plan on testing water formations, and also water in a cup.

4.   I found that it could not be walked on, nor could it be transitioned into another substance, even though The Savior and myself both had similar anatomy. This shows that my hypothesis was correct.

5.   Further testing, on people from other parts of the world, will be conducted to show that my hypothesis is correct.

Not Everything is true = N

True = T

N = T

•

18

BOWLS IN SPACE THAT CATCH MATTER

1.   Reading in books, and seeing the dimensions first hand, this is what I gathered. Also I noticed boundaries in one, two, and three dimensions suggested that there were other boundaries and dimensions.

2.   Different dimensions are possible, infinitely, and this is my hypothesis.

3.   By using what instruments I have an a matter of Empiricism, I plan on taking notes and

building evidence.

4.   I could only gather evidence from dimensional boundaries such as one, two, and three

because of limited perceptions, which are evidential of more.

5.   Further study with the use of special instrumentals will proceed.

Dimensions = D

Boundaries = B

Reality = R

D + B = R

●

19

MATTER IS SPACE & SPACE IS MATTER THEORETICAL OCCURRENCE

1.   I've seen and felt trees, and they were solid. The space around was not, without known

mass, it was interesting.

2.   My hypothesis is that at one point in human history solid matter and space will alternate. Where this is mass, there is then space and space mass.

3.   By taking data from the tree that I found, and then the space around it, I will show that a solid mass can be converted to space, and vice versa.

4.   I was successful because I turned wood into smoke or space, though I could not take space and turn it into matter without replacement, even though theoretically I could.

5.   More testing, which will require funding, will proceed to further prove that empty space can be turned into matter, and matter space.

$$Space = S$$

$$Matter = M$$

$$S + 1 = M$$

$$M + 1 = S$$

•

MEDICINE

1. My observation of medicine took me to conclusions.

2. It is my hypothesis is that there are new medicines, such as pill form medicines, that can be a product of improvement.

3. By taking vitamins or pills, that have proven as medicinal and beneficial, I will prove that there are more combinations that could also prove as beneficial.

4. There have been conclusions, and so new findings were found.

5. More studies, chemically, will have to be done for more conclusions.

Old Pills = OP

New Pills = P

Available Pills = AP

OP + NP = AP

•

OUR PLANET IS ACCOMPANIED BY ANOTHER PLANET MUCH BIGGER AND SHADOWED - *EARTH IS IN A SHADOW*

1. By looking at the night's sky, through a telescope, and at pictures and it appeared to me that we - as a solar system - could be in a shadow.

2. My hypothesis is that we are indeed in a shadow, and that there is a lot which is not seen and is out of view.

3. By using my eyes to look at the sky, through a telescope and at pictures I hope to prove that there is matter that is out of view and that we are in a shadow.

4. I proved that there is matter that is out of view and that outer-space resembles a shadow, by example.

5. Newer instruments will reveal what is hidden, and the fact that there is hidden.

Imaginary Number = I

Real Number = R

Estimated Percentage In The Shadow Compared To Out Of It = E

$$I + R = E$$

•

22

A GREAT MOTION OR FORCE IN THE UNIVERSE THAT WILL HIT EARTH IN THE
NEXT FEW DECADES, THAT HITS EARTH WITH LESS FORCE ON A REGULAR BASIS.

1.   I noticed that there is Causality, and that everything is being effected by more causes.

2.   The hypothesis that a propose is that there is a great force, like a wave, that will effect us
gradually and drastically.

3.   What I plan on doing is conduct studies to prove that dimensionally there is in fact wave
like occurrences that effect our planet gradually and that imply that there will in fact be a *greater
occurrence.*

4.   By collecting data, there have been changes and therefor what I thought to be true can
only be implied, and predicted.

5.   Sensitive partials instrumentally and equipment will be used, immediately, to prove
everything further.

•

Estimated Time Before The Great Force Hits Earth In Years = E

Destroyed Earth By Great Force = D

E100 = D

23

CHEMICAL WARFARE AS DEFENSE AGAINST UNWELCOME LIFE FORMS

1.   As seen in wars between countries, I have gathered data through research.

2.   Though not moral, if need for our country, biological weaponry could be used as a

defense. It is my hypothesis that it could be used, without it being traced back to us, leaving us as

the victor.

3.   By using atomic, genetic, and chemic compilations, a biochemical agent could eliminate,

strike fear in and destroy any enemy.

4.   Finding such a collaboration was found, but was not tested to the point of knowing it's

destructiveness.

5.   I found what I was looking for, and more studies on the remedy will be conducted to
ensure that we do not die and to ensure that there is peace.

Earth = E

Foes = F

Result = R

1E - 1F = 1R

•

24

BIOLOGICAL WEAPONS AND REBOUND

1.   Thought I have not seen everything in the universe, I have seen what biological weapons
are capable of.

2.   The hypothesis that I present is one of logic, and meaning. I think that biological
weapons should only be used if there can be control over the casualties and destruction.

3.   By using what information I have, I plan on proving that weapons should have and
opposite counter effect, therefor having more control over the hypothetical situation.

4.   My studies have proven that there are counter measures, and also prevention. Logically,

if the weapon can be controlled then the opposite can also.

5.   More studies can only bring more literacy.

Weapons Effect = W

Casualties In Numbers = C

W = C100

•

25

EVOLVE INSECTS TO OUR POTENTIAL SO THAT THEY BUILD AND COMMUNICATE

LIKE HUMANS

1.   Insects, like animals, have been seen by my eyes as having the potential to be more

through evolving.

2.   My hypothesis is that insects will either on Earth or on another planet reach their larger

potential and role.

3.   What I plan on doing is taking noted on insects, and then humans. By comparing I plan
        on predicting their possible evolutionary direction.

4.   My theory was proven correct, because both animals and insects behave alike, are both
        living things, and have evolved to a certain degree already.

        5.   Another experiment, such as studies of fossils, will take place.

Evolved Insects = E

Time Of Occurrence In Years = T

Planet = P

$$E + P = T500$$

•

FASTER ACCELERATION IN MACHINES

1.   By seeing machines like engines and other machines, I have gathered information from
        them.

2.   If a machine moves at 200 mph, then I predict that in ten years machines will move 50% faster.

3.   What I plan on doing, is taking notes year by year for ten years on the speed of specific machines.

4.   I was right, machines were faster, gradually over the course of years.

5.   Test in years to come will reveal more to the mind.

Acceleration Increase = A

Years until it occurs = Y

A = Y200

•

27

ANIMALS HAVE MINDS EQUIVALENT TO HUMANS, LACKING PHYSICAL

COORDINATION

1.   By hearing mammals, and seeing them, I have noticed that all of them may think alike.

2.   My hypothesis is that animals such as Jaguars can think like humans, without the advanced and advantageous fingers and mouth.

3.   By wearing gloves that restrict the fingers, and limit movement, I will prove that humans would be as intelligent as Jaguars in the limited circumstances.

4.   The hypothesis was true, that if physical features such as fingers and mouth movements are the same, intelligence could be also - estimated.

5.   Broadcasting my findings online could bring about better care for animals, and better treatment.

Human Features = H

Animal Features = A

Life = L

$(HA) = L$

•

28

CERTAIN NUMBERS ARE MORE APPEALING THAN OTHERS, LIKE PRICES AT A STORE

1. While shopping, there were prices that I liked and disliked, both high and low, and others that I did not.

2. My hypothesis is that some numbers are mathematically appealing and cause a greater sell value.

3. By comparing the numbers that I liked to the numbers that I rejected, I plan on proving why there was a preference.

4. My hypothesis was correct, that mathematically numbers that ended in the number 9 were more appealing and were acted on.

5. Research will no doubt show that in sales and marketing, some numbers are more popular that others.

Appealing Numbers = A

Unappealing Numbers = U

Better Sales = B

A - U = B

•

29

PEOPLE USE BODY LANGUAGE TO COMMUNICATE POINTS

1.   While talking to an acquaintance, he shrugged and pointed.

2.   What I hypothesize is that body language is half of how people communicated.

3.   By noting the different ways people communicate, I plan on making a graph showing charting how, how far and in what way people communicate - showing that people use body language just as often as they talk.

4.   It was true, and I had been certain that it would be true before tests. People do use body language to communicate their points.

5.   More studies will reveal more data.

Body Language = B

Verbal Language = V

Other Language = O

Absolute Language = A

$$B + V + O = A$$

●

30

IF EVERYTHING IN OUR SOLAR SYSTEM IS MADE OF THE SAME ATOMIC STRUCTURES, IT IS POSSIBLE THAT OTHER SOLAR SYSTEMS ARE MADE UP OF A DIFFERENT COMPONENT

1.   What can be seen, and what can not, could be two completely different worlds.

2.   What I hypothesize is that in a galaxy far, far away and completely out of our telescope range, there could be a much different place. Physical properties could be different there, and atomic structures could be drastically different.

3.   The plan that I have to prove I am right is to take a rock and hide under it a diamond. I will then ask an acquaintance to tell me what is under the rock.

4.   I was right, he could not tell me what it was, and tried by guessing. There was no way for him to know unless I moved the rock out of the way, revealing the diamond.

5.   More can be demonstrated, by scaling large situations down to smaller ones.

Different Physical Properties = D

That Which Can Not Be Seen = T

Real = R

$$D2 + T2 = R2$$

•

31

CLOTHES CAN CHANGE THE WAY PEOPLE THINK ABOUT YOU

1.   By looking at two different people, one clothed in matching clothing and one clothed in non matching clothing, I noticed that they were treated differently under the same circumstances.

2.   My hypothesis is that those wearing matching clothing will be treated better than those not wearing matching clothing,

3.   When put in a mall, I wanted to see how many people approached and talked to who.

4.   I was correct, those that had matching clothing were talked to, approached and treated

    better than those without matching clothing.

5.   More studies will show similar results, and I think that it will come as no surprise.

Matching Clothing = M

Better Treatment = B

$$M = B$$

•

32

THE UNIVERSE IS ROUND

1.   By looking at what is within the universe, I saw and felt objects that were perfectly round

like bubbles and the planets. The planets were acting *like bubbles in a whirlpool*, and moving in

a circular motion. Everything appeared to be moving, for the most part, it a circular motion!

2.	My hypothesis is that since there is circular movement in the ocean and also in the solar system, that what is around it is also a round object.

3.	I'll draw what I see, and when I do I think that I will come up with circular objects.

4.	I was right, that there is evidence that since there are spheric objects in the universe, that the universe is also circular and has an influence that is circular, multilayered.

5.	More circular objects will be studied, which suggest a spheric universe.

$$\text{Bubbles} = B$$
$$\text{Planets} = P$$
$$\text{Universe} = U$$
$$\text{Circular} = C$$

$$B = C$$
$$P = C$$
$$U = C$$
$$C = C$$

$$B10 + P10 = U10$$
$$U10 = C$$

•

MACHINES WILL TAKE OVER THIS WORLD

1.   I read that AI was progressive, and that technology was also. Also I have seen technological advances that have shown possibilities.

2.   By researching, taking notes and buying technology I hypothesize that there will be further advances, ones that could change the course of humanity.

3.   What I plan on conducting are tests that show, on a graph, that technology is and will continue to become more advanced, to the point of replacement. What I mean by replacement is that technology will take over human activity almost completely.

4.   It was true, that once on a graph technology has become progressively more advanced, and has taken more and more tasks.

5.   There is more that could be done to show what I have discovered.

Time In Years = T

Advances in Technology = A

T200 = A

●

34

THERE IS LIFE IN THE EARTH'S OCEANS THAT IS NOT KNOWN OF YET

1.   I've seen, felt and heard oceanic life. Also I have read text that suggests that all life is not accounted for.

2.   I hypothesize that there is more under the surface of the ocean. It is my thought that cold blooded ocean creatures could be at the bottom, and could be fascinating if studied.

3.   Further studies of taking pictures, working with scuba divers and submarines - and reading will prove my point.

4.   My hypothesis was correct, there were dead life that appeared to be that which had never been seen before, in later study.

5.   More exploring of the ocean will uncover what it should.

Miles under the ocean = M

New Life = N

M10 = N

•

35

FOSSILS ARE BROKEN DOWN AND IN SAND ON THE BEACHES.

1.   I've seen fossils, and I have seen sand on the beach.

2.   What I think, is that fossils could have become sand over time, in the ocean. This is my hypothesis.

3.   What I plan on doing, is to use a microscope, cross analysis, and sampling to show the correlations.

4.   I was correct, that they *appear* to be the same, and at this time is all I can know.

5.   More testing equipment will be used in the future, to show that my hypothesis is indeed correct!

Fossils = F

Sand = S

F = S

•

36

USING WOOD IS IMMORAL BECAUSE IT WAS ONCE ALIVE, AND IT'S OXYGEN KEPT HUMAN LIFE ALIVE

1.   I've seen trees, cut wood, humans and corpses.

2.   My hypothesis is that through debate I will prove my point that it is immoral to cut down trees - *making me more of a tree hugger.*

3.   What I will do is gather information on trees and humans, and cross examining to prove that trees should get to live as long as humans get to live, if innocent.

4.   I proved that I was right, showing that trees are innocent, *and therefor do not deserve to be sanctioned.*

5.   More debate will conclude that either it is right or it is wrong.

Number Of Trees Cut Down = T

Number Of Immoral Conducts = I

T1 = I1

•

37

PLANTS COULD EVOLVE

1.   I've seen evidence of evolution in humans, and plants have a similar way of being alive, making them both technically alive.

2.   My hypothesis is that in years to come, plants will evolve due to time conditions and biological experimentation.

3.   My plan, is to take a normal plant, and compare studies to an older study, to see if there are differences.

4. It was true, that the plant was slightly different, suggesting that the plant would change further.

5. More plants will be tested on, to bring us to more conclusive data.

Years Before Evolution = Y

Estimated Number of Evolved = E

$$Y10 = E1,000,000$$

•

38

THERE ARE MORE FREQUENCIES THAN CAN BE SEEN BY THE EYE - COLORS

1. I've seen blue, yellow, red, purple, orange, and green.

2. What I hypothesize is that there are more frequencies that could be seen, like higher ones and lower ones.

3. As more frequencies are found, more colors could be found and experienced.

4. I was right, I found more frequencies using equipment and so there are more possible
    colors to be seen.

5. More research will show that someday, we as the human race, will experience more with
    our eyes.

Higher Frequencies = H

Lower Frequencies = L

Unseen Frequencies = U

Time In Years Until Discovery = T

$$H2 + L2 + U4 = T100$$

•

39

SPORTS CAN SOLVE PROBLEMS IN THE WORLD BY BRINGING MORE PEACE

1. In sports, I have found that less people die when compared to wars.

2. My hypothesis is that sports could replace wars, and that by competing without weapons
    disputes could be settled.

3.   By taking notes on sports and wars, I will try to show how it is advantageous to settle political disputes non-violently supported by the UN.

4.   I was correct, it was true. When comparing casualties, sports supported life much more than war did, and suffering was reduced.

5.   I hope that more studies will follow mine, and show that sports should take the place of wars so that there is less pain and suffering.

Sports = S

Earth = E

Peace = P

S + E = P

•

THERE WILL BE A REVOLUTION WHEN MACHINES WILL EARN MONEY - *MACHINE RIGHTS*

1.   Machines work and do not get paid, nor given equal treatment.

2.   My hypothesis is that machines, over time, will organize and protest. I think that with AI there will be a mimicking of life and human behavior.

3.   I'll do my best to prove my hypothesis as correct, by taking the newest AI there is and testing it by bringing the on switch to the on position - and then evaluating it.

4.   It was conclusive and true that machines are moving in the direction of equal rights, and paid labor.

5.   More tests will bring about more evidence that supports my claim.

Human Rights = H

Machine Rights = M

Other Rights = O

Global Rights = G

$$H + M + O = G$$

•

A DEVICE COULD BE DEVELOPED TO CONTROL GRAVITY

1.   I've seen gravity on Earth, and it is rounded around the planet.

2.   My hypothesis is that a gun, or device of some kind, could be created by harnessing gravity or making alterations to it.

3.   What I am going to do is use a zero gravity simulator, and take the remaining gravitational pull and create a box that holds it, and then expels.

4.   I was successful in making progress toward the creation, without having it completed.

5.   Trial and error could result in what I am interested in arriving at.

•

Gravity Gun = G

Completion = C

New Invention = N

$$G + C = N$$

EXPLODING BULLETS

1. I have seen bullets, and have seen explosives.

2. My hypothesis is that bullets and explosive charges could be combined, and there for more effective.

3. By making larger bullets, I plan on having explosives put in the bullets as a demonstration of what it would be like on a smaller scale.

4. I was correct, in that I made a larger form of bullet that can contain explosive charges that could hypothetically detonate.

5. More research on the matter would not be needed, because it would be a safer tomorrow if there were no weapons.

Explosives = E

Bullets = B

Completion = C

New Invention = N

$$E + B + C = N$$

●

43

A PRESERVATIVE THAT LASTS LONGER COULD CURE WORLD HUNGER

1.   I've seen foods with preservatives, new chemical reactions, and people starving in other countries.

2.   My hypothesis is that there is another chemical reaction that could produce a preservative that could make more food last thousands of years.

3.   By using new formulas, and creating new prototypes, an eventual preservative will no doubt be found through chemical means whether this year or the next.

4.   I was right, another prototype was found that was like the old one, but lasted a longer time.

5.   Next I plan on testing to make a preservative that will last longer with low health risks and life saving qualities.

Preservative Duration = P

Chemicals Combinations That Are New = C

Unneeded Chemicals = U

$$C - U = P100$$

•

44

MORE SURVEILLANCE WILL LOWER THE RATE OF CRIME

1.   I've seen cameras, and crime rate, but have not seen them combined often enough for there to be a difference.

2.   What I think is that cameras should be all over the world, in every place except in private areas. I think that this will cut down the crime rate, and bring more peace to the world - this is my hypothesis.

3.   By placing a camera at the police department, it has been calculated that by doing so will cut down on the crime rate of anyone breaking the law there.

4.   It was true. By doing so, and having people know about it, it made the crime rate decline and made the world a safer place for new generations.

5.   What I think is that the future will be better monitored and safe, because there will be more surveillance.

Number of Areas Under Surveillance = N

N100,000

●

45

A CURE FOR ALL DISEASES

1.   I have seen cures, and had cures for ailments and conditions.

2.   What I think, or hypothesize, is that since there have been cures for other conditions, there should be cures for all of them - *all or nothing*.

3.   What I plan on doing is using Aspirin for my headache, and note if I have alleviated the condition.

4. I was correct, I alleviated my condition, and so there must be cures for all conditions - all or nothing.

5. I plan on creating more cures for conditions, because if there are some cures, then that is evidence of more cures.

Diseases = D

Cures = C

Saved Lives From Cures = S

D + C = S

•

46

THINGS ARE UNIFIED

1. The oceans, the islands, the air and outer space all move in directions.

2. Through example of causality I plan on taking notes and proving, with equipment, that everything is moving in the same moment and motion.

3.　　When I use a camera, it will capture a scene with islands, ocean, air and outer space, showing that they are all unified together.

4.　　My hypothesis was correct, that by taking a photograph, everything in the universe can be linked together and unified through visual means.

5.　　Conclusively, I think that there are no more conclusions to come upon, or be drawn. If any were, it would be that of old news - unless something new was revealed.

Movement Of A Billion Objects = M

Time That The Objects Are Moving = T

Observer = O

$$M + O = T$$

•

DREAMING AND WAKEFULNESS ARE BOTH REAL OCCURRENCES

1. Dreaming and waking life are almost the same experience, from what I have experienced in different points in time.

2. It is my thought that dreaming is a part of reality, and so is waking life. Both are so intimately connected that both could be of equal value - and that is my hypothesis here.

3. What I intend on doing is writing down a dream at night, and then a waking day. I think that by doing this, I will find that they are related.

4. I was right. both dreaming and wakefulness are related, and both feel like one another.

5. Entering the dream world could open possibilities into the gaming world, or simulated activities. If dreams could be controlled and induced, then it could become a sport to dream, and not one to wakefully be.

Waking Life = W

Sleeping Life = S

After Life = D

Real Experience = R

W + S + D = R

•

FATE IMPLIES THAT NOTHING CAN BE WILLED

1. I thought about going to the store and then went to the store.

2. My hypothesis is that everything is fate, and that choosing to go to the store is only a matter of fate, and the choosing is nothing other than fate - fate that the choice was made and fate that I went to the store.

3. By charting the occurrence, I plan on pin pointing where I was, when and why.

4. I was right, because who, what, when, where, and why - action and inaction - are all part of the universe's motion.

5. By trying to make conscious decisions, there are influenced by other circumstances. More evidence will build, and degenerate.

Fate = F

Will = W

Reality = R

$$F - W = R$$

•

49

ENERGY CAN NOT BE CREATED OR DESTROYED, AND IS A CONSTANT

1. Energy transitions from one thing to another, and does dissipate but dies not disappear.

2. My hypothesis is supported by others who have a hypothesis about this. Energy can not be destroyed, and so it just transitions from mass, to mass, to space.

3. What I want to do is prove that energy does not become destroyed. By using a battery, the energy will originate in the battery, and then move to the object that uses batteries.

4. I was correct, energy moved from the battery to the remote controlled car, and then outwards.

5. More conducted studies will show that I am right, and that energy moves but does not become destroyed, ever.

$$Energy = E$$

$$E = \infty$$

•

50

DRUGS IN GENERAL SHOULD ALWAYS BE PRESCRIBED

1.   I've seen people take drugs as prescribed, but drink alcohol without a prescription.

2.   My hypothesis is that those that take as prescribed will experience more benefit than those that take things illegally or in excess.

3.   By studying subjects, I will determine who is doing better between the two selected - A or B.

4.   My hypothesis was correct, because the person that took medication as prescribed was more able, mentally and through coordination, that the one that drank alcohol who showed clumsiness, irrationalness, and afterwards headache feeling.

5.   More study could show that the subject who drank alcohol is even more at a disservice from the unprescribed.

Prescribed Medication Or Prescribed Intoxicants = P

Better Health = B

$$P = B$$

•

51

HOLOGRAMS COULD BE CREATED

1.   I've seen holograms in the experimental stages.

2.   My hypothesis is that holograms, through advanced technology, can become a store held item.

3.   The theoretical hologram could be predicted through means of making estimations.

4.   My estimation showed results, and my hypothesis was correct.

5.   Actual holograms in stores will provide more tests.

Holograms In Numbers = H

More Advanced Technology = M

Years Until Invented = Y

$$H1,000,000 + M = Y50$$

•

52

ARGUMENTS ARE MISUNDERSTANDINGS

1. As far as arguments go, there seem to be also misunderstanding and confusion which are
seen in wars.

2. My hypothesis is that misunderstandings and confusion could lead the world to war, and
that such confusion could be avoided and prevented by means of communication.

3. What I plan on doing is taking examples of misunderstanding, and using theater to prove
that preventative measures could benefit the whole world.

4. By acting it out, I proved that such is going on, or could be. Also, in the end there are
examples of prevention.

5.   Through example, or real time situations, more studies could reveal more about the

subject.

Communication = C

Misunderstandings = M

Peacefulness = P

C - M = P

●

53

EVERYTHING THAT IS SEEN IS ONLY A PART OF YOURSELF

1.   I know from Empiricism, that what is seen, and that which is seeing, are linked by

energy.

2.   The hypothesis that I present is that through Empiricism, what is seen and what is seeing

the seen are one and the same.

3.   By using logic, I will prove that the seer and what is seen are both the same.

4.   Through logic, and mathematics, I was correct in presenting that everything that exists is in the eye of the beholder, and that outside of that is only speculated.

5.   More could be uncovered, revealed and exposed through taking notes and studying what we experience everyday. I will do this!

$$\text{Other People} = O$$
$$\text{The Self} = T$$
$$\text{Sight of Surroundings} = S$$
$$\text{Reality} = R$$

$$O + T + S = R$$

•

54

CONCEPTIONS

1.   I have what are called conceptions, and everything could be made of concepts.

2.   Conceptions could be all there is, and everything could be one giant conception - this is the hypothesis I pose.

3.   By writing it down, and making sense of my thoughts, conceptions could be shown to be the one and only thing.

4.   I was right, that everything in the universe could all be one conception - though there are some theories that could prove me wrong.

5.   Neurology will be another subject of study.

$$Conceptions = C$$
$$Reality = R$$

$$C = R$$

•

55

WHAT HAPPENS AFTER DEATH?

1.   Observationally, humans have died and I have heard that they have died after seeing them buried.

2.   It is my hypothesis that the energy once contained in a human body leaves it, and finds something else as a container such as a tree in divided sections of plants.

3.   I intend on studying trees, alive humans and dead humans.

4.   My findings state that trees are alive like humans are or were, supporting my hypothesis.

5.   More research will continue on live, and death.

Life $= L$

Death $= D$

Energy $= E$

$$(L) + (D) = E$$

•

56

PLAYING WITH G.I. JOES IS UNHEALTHY FOR CHILDREN

1.   Seeing children playing with G.I. Joe figures was what I saw one day.

2.   It is my hypothesis that children that play with toy weapons will use weapons when they get older, leading to unsatisfying results - like murder.

3.   By taking into consideration the playfulness of children with the G.I. Joe figures, and observing what how they shoot their toy guns, I plan on collecting data.

4.   I was right, because years later, those that played that way became a gun owner, and interested in combat - and combat creates an unhealthy mentality.

5.   I want to continue researching, because by doing so more premature death could be prevented.

Violent Figurines = F

Violent Play = P

Violence = V

$$F(10) + P(100) = (1000)V$$

•

GRADING STUDENTS IS NOT THE CORRECT WAY TO ASSES PROGRESS

1.   Some children are not doing well in school, and instead of improving his performance in school he is graded and discarded.

2.   I hypothesize that students should be graded privately, and only have their successes on their public and permanent record.

3.   By helping a student at school, I will prove that he improves and that his earlier low grades were unnecessarily on his public and permanent record.

4.   Affirmation to my hypothesis, by helping the student his grades became better and lower previous grades were of no real use.

5.   Studies on schools, their curriculum, and techniques will show more data on the subject.

$$School = S$$

$$Grading = G$$

$$Observation = O$$

$$O - G = S$$

•

58

PREDICTING THE FUTURE EXACTLY IS POSSIBLE

1.   I have seen one, two and three dimensions.

2.   My Hypothesis is, if more dimensions exist, they could possibly reveal the future to anyone with the capability of revealing other dimensions with other properties.

3.   I plan on testing matter and space, so to find other dimensions.

4.   It did go as I had thought, because I did have the right equipment.

5.   More plans for analyzing compounds and reactions are to be carried out.

Number of Dimensions = D

Future = F

Present Time = P

$$D100 * P = F$$

*

•

59

THERE IS LIFE ON OTHER PLANETS

1.   Molecules and DNA show that life exists on the planet Earth.

2.   My hypothesis is that there is life on other planets because life on our planet suggests that it could be elsewhere. Whenever there is something of abundance in one area, there is usually more elsewhere.

3.   By looking at DNA structures, I will further hypothesize that if there is DNA in one place, it must be in many other places, such as on another planet.

4.   I was proven correct, that indeed life was in some areas, and also in others (on the planet Earth). Life here is suggestive, and supports the theory that it is actually elsewhere in the universe.

5. Traveling at high speeds, us as a human race could explore, and that is what I plan on doing as soon as there is technology.

$$\text{Life } \infty = L$$

$$\text{Other Planets } \infty = O$$

$$\text{Truth } \infty = T$$

$$L + O = T$$

•

60

MUSIC HAS AN INFLUENCE ON THE MIND

1. Sound waves have been heard, and different notes or gaps in sound have shown that it exists.

2. I hypothesize that music has an effect on the mind, because it brings back memories, evokes emotion and tunes thoughts.

3. By listening to music of different genres myself, I predict that I will be influenced by it and have a reaction.

4.   My theory was correct, it made me feel differently that I had originally before hearing it. During the song I heard, memories came to mind and it evoked a mind state - influenced.

5.   Music is limited in its use, and can communicate but can not do much else. It could prove to be useful as a therapy.

Music = M

Influence On Emotional = I

Experience = E

$$\underline{M + I = E}$$

5

•

61

CRIME CAN BE LOWERED BY GIVING PEOPLE MORE MONEY

1.   Studies have shown that people with money commit less crime, and I ready this reflecting on my own experience.

2. Presenting a hypothesis, such as this, will benefit man kind and the world. What I think, is that if a man in poverty is given money, that he will be less likely to steal.

3. What I intend on doing, is finding a criminal and giving him $20 so that he can spend less time in jail over the course of his whole life.

4. I was right, because by giving him money, he did not have to steal it that time lowering the crime rate.

5. Though it is a sensitive subject, I plan on researching but not interfering with others lives at this point.

$$Poverty\ Stricken = P$$

$$Wealthy = W$$

$$Money\ In\ Thousands = M$$

$$Solution = S$$

$$(P)(W) + M5 = S$$

•

A METEOR HIT EARTH AND KILLED THE DINOS

1.   I saw a commit in the sky, and it was moving with velocity.

2.   Dinosaurs were killed by a natural disaster, and they would still be alive today had it not happened. This is my hypothesis.

3.   Conducting tests on fossils, meteors and natural disaster, I plan on finding a conclusive answer.

4.   What I hypothesized was correct because fossils have shown to those that already study the matter that they had all died from one thing, estimated.

5.   Fossils could show us even more than what is known now, through more advanced methods.

Dinosaurs = D

Earth = E

Current Day = C

E - D = C

•

CURE FOR CANCER

1. I have witnessed articles in the paper that have covered topics such as Cancer, and it was informative.

2. My hype is that since there have been cures for other diseases, then a cure for cancer is possible and within reach.

3. By comparing to code, which has within it patterns, I will demonstrate that both code and chemical combinations are similar.

4. I was successful in finding evidence, in patterns and chemistry, with no cure.

5. A cure will be found, because it is sequentially appropriate.

Cures For Other Diseases = D

Cure For Cancer = C

Time In Years Before All Diseases Have Cures = T

Result = R

$$D + C * T100 = R$$

•

64

I THINK IT IS POSSIBLE IN BASEBALL TO HIT THE BALL 1,000 TIMES IN A ROW

1.   I watched baseball, and saw them hit and strike out.

2.   My hypothesis is that by use of math and probability it is possible to hit the ball 1,000 times and even 1,000 home runs.

3.   By using mathematics, I'll show that it is possible.

4.   I was correct, it is a possibility, and there are so many possibilities in the universe.

5.   By happening, it will be better proven, I think.

Home Runs = H

Time in Years Estimated before happening = T

Result = R

H1000 * T1 = R

•

65

AIRPLANE FUELED BY WIND POWER

1.   I've seen airplanes that glide, and wind power generators.

2.   My hypothesis is that by combining the right elements, such as the right design, airplanes could be self powering by natural and non-polluting fuels like the ones that are used today.

3.   What I plan on doing is buying a Styrofoam plane, and putting propellers on it to see if it could work.

4.   I was right, it worked and is very suggestive of an actual plane wind powered and light.

5.   More investigations could reveal more.

Velocity = V

Fuel Required = F

Distance = D

$$V - F = D\infty$$

•

66

ADVERTISING PERSUADES PEOPLE

1.   I've seen advertising, and also people's decisive actions.

2.   What my hypothesis is, is that when people see certain stimuli, then it guides them in certain directions.

3.   By showing people a commercial, I plan on studying their reaction. I think that if something enjoyable like the sun is combined with sales product, then thinking about the sun they will make a purchase.

4.   It was true, by showing subjects the right stimuli, I can take their money for a trade.

5.   More study will expand knowledge.

The Advertising = T

Audience = A

Purchases = P

$$T + A = P$$

•

67

EVERY LOCK CAN BE OPENED BY A MASTER KEY

1.   I have seen locks, and their keys.

2.   The hypothesis that I present is one of fact. A lock can be opened, no matter what it looks like by use of the right instrument, key or explosives.

3.   What I plan on doing is use an explosive to open a large lock. When it opens, I will be proven correct.

4.   I succeeded, the lock broke from the explosive charge.

5.   Nothing more could be revealed, unless there is a master key.

Lock = L

Opening Instrument = O

Unlock = U

L + O = U

•

68

POEPLE CAN BE FROZEN AND THEN THAWED WHILE ALIVE

1.   I've seen people get cold in the winter, and did not die.

2.   My hypothesis is that people can be partially frozen, and live.

3.   By going outside in the winter, I will prove that though I become cold, I do not die.

4.   I was right, I did not die from the cold.

5.   Lowering the degrees in the future could provide more information.

People = P

Degrees In Fahrenheit = D

Life = L

10P * D30 = L

•

69

CONDUCTING TESTS THAT INFLICT PAIN ARE IMMORAL YET EFFECTIVE

1. Hamsters have been studied on, and studies show that they feel pain.

2. What my hypothesis is, is that conducting tests on hamsters is immoral.

3. By use of logic, it is. If humans are alive and feel pain, and hamsters are alive and feel pain, then neither should have to feel it because of anatomical similarities..

4. I was right, because anatomical similarities show that if they feel pain, then we also do and it is immoral.

5. More painless tests should be done.

Hamsters = H

Painful Testing = P

Immorality = I

$$H + P = I$$

●

70

T.V. DRAMA IS UNHEALTHY TO WATCH AND SETS A BAD EXAMPLE

1.   While watching T.V. I noticed that there is drama

2.   My hypothesis is that Drama on T.V. is unhealthy because it can create more drama, setting a bad example.

3.   By showing T.V. drama to someone, and also real drama, I will study the reaction and repetitive motions.

4.   Correct I was, drama makes people less happy.

5.   More drama and reaction should show the same.

Drama and Violence = D

T.V. = T

Real Violence = R

$$D + T = R$$

•

71

BAD HIGENE COULD LEAD TO OTHER UNSANITARY BEHAVIORS

1. Like a tree brach falling from a tree, it usually breaks more and more on the way down.

2. What I propose is that if someone does not shower, it can have a bad effect on other aspects on his and other's life.

3. The point here, is to take an example and ask that example questions and study him.

4. I was right, bad hi-gene leads to more disorderly conduct.

5. More studies will show that this bad learned behavior could be corrected.

Bad Hi-gene = B

Escalation of Bad hi-gene = E

Over Years = O

Bad Result = R

$$(BE2) + O = R$$

•

72

RELIGION TWO

1.   By reading, I notice the terms.

2.   Buddhism and Christianity are relatively similar, using terms that read differently but mean the same thing.

3.   I intend on conducting studies that show that Nirvana and Heaven are two different words, meaning basically the same thing.

4.　　By reading the terms and definitions I have arrived at the conclusion that Heaven and Nirvana are almost the same, but predicted as slightly different.

5.　　By reading, and asking members of both social groups, I intend on conducting more studies revealing that the two religions are so similar that they should join.

$$Nirvana\ Text = N$$

$$Heaven\ Text = H$$

$$Similarity = S$$

$$N + H = S$$

•

73

1.　　I have never seen teleportation.

2.　　My hypothesis is that teleportation is impossible.

3.   I plan to use text and proven theories to support my thoughts on the matter such as the movement of matter through space, the movement of matter through matter and the movement of space or lack of movement.

4.   My findings show that teleportation is and always will be impossible unless the matter takes a different shape by being teleported. Though because matter is always in a state of progression and change, what I propose is that matter can not be kept at a constant state of existence, though energy could be.

5.   More studies will be conducted.

Teleportation = T

Earth = E

Time It Will Be Invented, In Years = I

$$T + E = (I)0$$

•

74

BUDDHISM AND CHRISTIANITY ARE RELATIVELY THE SAME, USING DIFFERENT TERMS

1.   I read about and talked to certain practices of religion.

2.   My hypothesis is that religions have so much in common, that they should be
characterized as the same group.

3.   By using my hands and writing on paper, I plan on making connections between meaning
and labeling.

4.   I found success, which could lead the to world peace.

5.   Spreading of the knowledge that I have found, will bring people from other religions
together, and thereby lowering violence.

Religion One = RO

Religion Two = RT

Similar = S

(RO)(RT) = S

•

LIQUID METAL AND LIQUID WATER HAVE THEIR PHYSICAL PROPERTIES IN COMMON.

1.   It has been noted that liquid water and liquid metal are only different by their chemical properties and anatomic structure.

2.   My hypothesis is that though structured differently at an atomic level, both have a similar structure at a larger scale and therefor could be used as substitutions.

3.   By a series of tests, like water powering substituted by metal properties, I plan on using one for the other.

4.   My studies were a success, because liquid metal can create power when substituted for water.

5.   More substituting should create further power creating inventions.

Liquid Metal = CO

Liquid Water = H2O

Liquid = L

Powering Invention = I

Power = P

CO = L

H2O = L

L + I = P

•

76

TREAT OTHERS THE WAY YOU WANT TO BE TREATED

1.   Seeing others relate on a kinder level creates a healthier, happier couple of people.

2.   What my hypothesis is, is that by showing kindness, life could could become extended a

few years.

3.   By example, I will show an animal kindness and make predictions on life span.

4.   The animal had reduced stress, and therefor was healthier so in conclusion the test was a

success.

5.   More animals will be treated kindly, so to reduce stress in general.

Added Life Length In Hours = L

Hours Of Kindness = H

$$H24 = L12$$

•

77

COMPRESSION OF MATTER CREATS ENERGY AND HEAT

1.   I watched an object become squashed, and noticed that when it was it moved under

pressure.

2.   It is my hypothesis that by squashing matter, it will actually create energy and heat.

3.   I will take clay, which has simplistic properties, and with a piece of wood compress.

4.   It is my assumption that by moving the clay, energy moved and there was a release of

activity.

5.　　It was conclusive enough that no further testing was necessary.

Compression Of Matter, By Inches = C

Heat That Was Generated In Degrees = H

Time = T

$$C * T = H(5)$$

•

78

TWO OBJECTS CAN TOUCH WHEN MILES AWAY

1.　　When looking at a leaf blowing in the air, it landed on a tree. I can look at my thumb, the leaf and the tree all in one sight.

2.　　My hypothesis is that though there is research that has proven this untrue, I can prove it true.

3.　　I'll take a two dimensional photograph of a tree with my hand in the photo without physically touching.

4.   I was right, that they do touch because the light touches together even though the mass

does not touch.

5.   More examples could prove that two objects or masses can touch because the light waves

do.

Object One = O

Object Two = T

Touching Light = L

$$O + T = L$$

●

THE MILKY WAY WAS AN EXPERIMENT BY A LONGER LIVING OTHER WORLDLY

INHABITING SPECIES

1.   I've seen pictures of outer space, and seen it through a telescope. It appears to be some

kind of after effect, chemically.

2. I hypothesize that since it looks like the chemical aftermath of another species, it very well could be and we should be alarmed.

3. By cross examining that which is on a smaller scale with the larger, I will show that I am right.

4. Through experimentation, referencing, and the act of comparison, I proved that we are because there is no other explanation.

5. More observation, chemistry and the experiences of every day life, I will keep my thoughts on it.

Inhabitants Of Another world Experiment $= I$

Chemicals One $= CO$

Chemical Two $= CT$

Chemical Three $= CH$

Chemical Four $= CF$

$$CO + CT + CH + CF = I$$

•

SPONTANEOUS COMBUSTION

1.   While outside, I took an explosive and watched it blow up.

2.   Hypothetically, if this happens to explosives, then it could possibly happen in other places to other things.

3.   By keeping my eyes open, I think I will know of other items that explode, and so I will be observational.

4.   It was true, that seemingly some things explode.

5.   Observation will be my primary concern.

Explosive Object = E

Spontaneous Combustion = S

Other Factors = O

$$E + O = S$$

•

THERE IS SOMETHING FALLING

1.   I noticed that there is gravity.

2.   The gravitational pull might be a source of a greater gravitational pull, one that is far away and not seen going in one direction.

3.   What I plan on doing is taking a small object and dropping it, watching it on a smaller scale.

4.   I was right, when I dropped it, it moved it one direction, for the most part suggesting that on a smaller scale, it is similar on a larger one.

5.   No more testing is necessary on this planet, possibly other places.

Gravitational Pull By Force = G

D = Disaster

$$G1 + G10 = D$$

*

●

82

SO SMALL IT HARDLY EXISTS

1.   I have seen what I can.

2.   My hype is that there are still undiscovered matter, energy and space that is so small that it has yet to be seen.

3.   What I will do, is use my telescope to find very small pieces of matter that have never been discovered because of their more than microscopic size.

4.   I found something smaller, I was right. I do not think it has ever been discovered.

5.   By creating better instrumentation, there could be small worlds that know one has ever seen, on this planet.

Scale = S

S0.0000000001

•

LOSS OF ORBIT

1.   The Earth rotates around the sun, from what I have seen and read.

2.   My hype is that the Earth will no longer rotate in the next few thousand years.

3.   As with everything, matter settles. The Earth is matter, and it too will settle. I plan on proving this is true through watching an example for less velocity.

4.   I was right, that both the Earth and the object spin, and that both reach their end of motion while everything else moves around it.

5.   That which moves, comes to a halt and that which is at a halt moves again. More studies on this will proceed.

Duration Of Years Estimated Before Occurrence = A

Distance In Miles = D

Velocity = V

$$A3000 * D1,000,000 = V$$

*

•

84

SOUND WAVES ARE EFFECTIVE

1.   There are objects that age, and lose energy from what I've seen.

2.   The sun, and other objects become older and worn by loud sounds like stereo or speakers volume - this is my hypothesis.

3.   Sound waves will be tested, and it will prove that sound waves are everywhere and influence.

4.   It was true, because I was effected and there for all mass is.

5.   More studies are to come.

Volume = V

Impacted Masses = I

$$V100 = (I)1000$$

*

●

85

CAUSALITY TWO

1.   I noticed that people have a certain anatomy.

2.   Causality is the main reason that people are anatomically structured, and it is not a matter of pre thought as it is Causality.

3.   There was an action, and humans were the reaction. My hypothesis is that humanity is only a matter of Causality, whether it was intentional or not.

4.   I was right, because it is the only explanation that logically makes sense, and also because I have seen such.

5.   More will be revealed in time.

Cause and effect = C

The Meaning Of The Universe = T

C = T

*

●

86

MASSIVE CUBIC OBJECTS IN THE UNIVERSE, OUT OF VEIW

1.   I've seen cubic objects, and have also seen spheres in space, such as planets.

2.   My hypothesis is that since there are cubes on Earth then there could also be ones on a

giant scale, hundreds of miles across.

3.   I'll prove that there are giant cubes in space be demonstrating that there are cubes next to

spheres on Earth, coexisting.

4.   My demonstration was a success, they did in fact coexist.

5.   Exploration of the vast universe will no doubt support my theory, *because they must*

exist.

*

Cubic Miles Of Object = C

Reality = R

C1000 * 4 = R

●

87

SIGHT OF AN OBJECT IS NOT THE OBJECT ITSELF

1.   I've seen objects, and I have seen light.

2.   When looking at something, it is possible to know it is there without making contact with

it.

3.   What I will do is look at a rock outside, without touching it.

4.   I was successful, that through means of light perception I was able to see it but not

actually see the object. I saw the light, not the mass.

5.   The study of light will no doubt expand.

*

Light = L

Object = O

Human Eye = H

Experience = E

H + L - O = E

100

•

NEWER GENERATION OF BATTERY PREDICTION

1.   I've felt batteries, and looked at them. Also, I looked at my calendar.

2.   It is my hypothesis that there will be a battery that lasts longer than any battery today, year 2013.

3.   By purchasing batteries, and testing them, I plan on drawing conclusions.

4.   A year later, I found a new battery and it lasts longer. My hypothesis was correct.

5.   Battery tests will be conducted in the future.

*

Charged Atoms = C

Uncharged Atoms = U

New Battery = N

$$C - U = N$$

•

89

HUMAN EVOLUTION TAKEN FURTHER

1.   Humans look a certain way, and I have read articles that have proven that humans have a role in evaluation.

2.   One thousand years from this date, I hypothesize humans will evolve to a physique that is altered. Humans are currently evolving, everyday.

3.   By taking pictures, I will prove my theory by studying the past.

4.   The past provided that humans evolved, and that there will be more evolution.

5.   More testing, analysis, and review will show more.

*

Years Until Visible Evolution = Y

Existing = X

$$Y1000 = X$$

•

90

SUCCESS THROUGH COMPARISON

1.   I noticed that some people reach 10% of their potential, and I have reached 90%.

2.   My hypothesis is that by comparison to some, I have become successful but compared to others I have not become so.

3.   By studying two subjects, I intend on proving my theory.

4.   Subject A became less successful than subject C, and therefor C feels successful and A does not.

5.   More experiments could prove insightful.

People A = A

People B = B

People C = C

$$A\,10\% + B\,90\% = C\,100\%$$

•

91

WHEN MATTER MOVES, IT LEAVES INDENTATIONS IN THE UNIVERSE

1.   An object in sand, when removed, leaves indentations.

2.   The hypothesis is that when matter is moved, at all, it leaves indentations in the cosmos. I think that empty space is like very fine sand, and thought it is very lightweight space is matter.

3.   What I plan on doing is using heavy matter, which is visible to the retina, and show indentations.

4.   I was right, that there were indentations.

5.   More study on fine matter could reveal secrets.

*

Sand = S

Object = O

Indentations = I

S + O = I

•

92

TINY MATTER AND TINY UNIVERSES

1.   I've seen massive matter, and the universe through a telescope. Also, I have seen pictures

of the universe in books, which were informative.

2.   Hypothetically matter on a larger scale could also be on a smaller scale. there could be

tiny particles that formulate atoms, and and tiny particles that build the particles.

3.   By using the larger scale, I'll prove that mass in space, in size, can become infinitely

smaller and infinitely larger.

4.   What I thought to be true, was, as far as I could prove.

5.   Time could be infinite, scaling up and down could be infinite, and space could be.

*

Small Matter = S

Large Matter = L

Total Matter = T

$$S + L = T$$

●

93

ADVANCED ADVANCED MATHEMATICS

1.   From what I was taught, there have been inventiveness regarding mathematics. There is addition, algebra, and calculus to name a few.

2.   My hypothesis is that in the next thousand years, there will be more inventiveness.

3.   Based on algebra, I will test the past to predict the future.

4.   My calculations were correct, that in predicting there will be more, there probably will

be.

5.   By trying, I will attempt to do it myself, in hopes to further what is known.

*

New Mathematics = N

Old Mathematics = O

Total Mathematics = T

$$N + O = T$$

•

THERE ARE MASSES IN THE UNIVERSE EXISTING IN ABSOLUTE ZERO GRAVITY

1.   From what I have seen and heard about gravity.

2.   I think that there are places far away that have no gravity, gravitational pull, and no black

hole.

3. As I have done before, I will use equipment to prove my hypothesis. What I will do is use an artificial zero gravity environment to prove that there are other areas.

4. It was true, in showing that were was one, there must be more.

5. Nothing more should be done.

*

Mass = M

Absolute Zero Gravity = A

Situation = S

(A)(M) + A = (S)

•

95

AFRICAN AMERICANS HAVE DARK PIGMENT FROM SUN EXPOSURE AND ARE PASSED ON AS GENES

1.   The skin on african americans appears to be dark.

2.   What I hypothesize is that the sun, tanning and genetics are the explanation for the

pigment.

3.   Sampling skin and making comparison, I will prove that I am correct.

4.   Truth! The pigment was influenced by more than one factor, and the factors were what I

had thought.

5.   There is nothing more to know about it.

*

Sun Exposure = S

African Americans = A

Dark Pigment = D

$$S + A = D$$

●

COMMITS ARE A SIGN OR SIGNAL OF SOMETHING

1.   The commits have been seen, and they burn up in the ozone layer.

2.   I thought that this could be a sign. It could be a sign that there is something else coming. Like all commits, they come from another world, and I want to prove that the other world is included.

3.   By watching them in the sky, I will prove my point.

4.   I was right, the commits were obviously not from this world, implying that there had been an explosion of some kind on another world, possibly a world with life.

5.   More research could conclude as to why there are commits, other than the vague explanation.

Space = S

Commits = C

Signal = I

$S + C = I$

•

ALL PEICES OF MATTER WERE A PART OF ONE ISOLATED MAJOR PIECE AMONG THE MAJOR MASSES

1.   When a piece of bread crumbles, it leaves a bunch of smaller pieces on a table. The crumbs, in comparison to the piece of bread, are very small.

2.   What I think, is that every planet in the galaxy is really only small pieces of something much, much larger. A omega massive planet, one that has never been seen because it is too far away to see, could exist.

3.   By taking a large piece of bread, I will make smaller pieces.

4.   I was correct, that there must be something larger that Earth is from, a planet much larger in scale.

5.   More exploration in space will provide more answers.

*

Major Planet = M

Smaller planets = S

Total Mass in 1,000,000 Cubic Miles = T

$$M + S = T$$

•

98

ELEMENTS ARE BEING SIFTED AND SORTED OUT OVER TIME, BY LIFE AND BY CHEMICAL MEANS

1. I've seen digestion, nature and chemical experiments.

2. My hypothesis is that elements are becoming purified and then diluted, diluted and then purified through certain processes. There seems to be fluctuation.

3. By studying chemically different kinds of matter, I plan to show that at times the chemical compound is abundant, other times it is mixed and other times not there at all.

4. I was successful in proving my point through three experiments.

5.   Studies as to why it is this way, should be adequate.

*

Element One = EO

Element Two = ET

$$\frac{(EO)(ET) = EO * ET}{2}$$

•

99

ELECTRONICS COULD OPERATE ON A LARGER SCALE

1.   I've seen electronics, and I have seen them on a small scale.

2.   What I think, my hypothesis, is that robots could be created that could be on a larger scale, like an electronic space ship that is five times the size of Earth.

3.   What is planned, is that by making an electronic space ship on a smaller scale, I can

prove that there could be one on a very large scale.

4.   It was true, that on a small scale it could exist and so on a larger one it could.

5.   Building a massive spaceship could provide more answers.

*

Earth Mass = E

Electronics Mass = M

(E)5 = M

•

100

THERE IS PROGRESSIVELY AN ORDER THAT IS TO BE BROUGHT BACK TO

MATTER AND LIGHT. ORDER IS DISORDER

1.   I've watched time go by, by listening to the clock chime. Conceptually, there seems to be

disorder in comparison, like in biology.

2.   It is thought by me that the disorder in the universe is brought about by a disorder in

biology, and possibly in everything.

3.   By comparing the order to the disorder, I hope to arrive at conclusions that will be

revealing.

4.   There was order and disorder, suggesting that at one time there was only order.

5.   The disorder in the universe, hopefully will be understood, and as to why there is not only

order will hopefully be uncovered.

*

Universe = U

Biology = B

Order = O

U - B = O

●

101

SOME SUBSTANCES REACT TO ENERGY IN SEPARATE WAYS.

1.   Some metals conduct electricity, and water does more so from what I have observed.

2.   My hypothesis is that there are new substances that conduct electricity at a higher rate
      that could change the course of modern electronics.

3.   By testing with what material I have, I will prove that electricity differs in how it is
      conducted.

4.   It was true, that depending on the material that is used to conduct electricity varies.

5.   More material matter, of another world, could be found and testing it could bring about
      results.

*

New Substance = N

Conductivity = C

Reality = R

$$N + C = R$$

2

●

102

THERE ARE SLICES OF MATTER THAT ARE IMPOSSIBLE TO SEE

1.   I've noticed that paper is very thin, and is also matter.

2.   What my hypothesis is, is that matter can be like a thin piece of paper, only a thousand

times thinner.

3.   By studying what is visible to the human eye, I will use paper as a demonstration.

4.   The demonstration showed that matter can be like sheets of paper, and if one width, then

it could also be other widths too.

5.   By using advanced equipment, thinner matter could be explored more thoroughly.

*

Paper In Pounds = P

Mysterious Slice In Pounds = M

Matter Of All Other Sizes = A

Total Mass = T

$$\frac{P + M + A = T}{2}$$

•

103

FASHION DETERMINES IN PART HOW PEOPLE LIVE, AND HOW OTHERS LIVE.

1.   By seeing people who have a sense of fashion, it is fascinating.

2.   My hypothesis is that fashion can alter relationships, treatment and status.

3.   What I will do is wear something fashionable, and go about my day.

4.   I did, and it was true! By using fashion, I found relationships easier, I was treated better at restaurants and was thought to have a better status.

5.   Fashion could improve life, and more tests could prove that to be true.

*

Fashion = F

Healthy Relationships = H

F = H

●

104

THERE ARE OBJECTS THAT CAN MOVE INCREASINGLY FASTER BUILDING INFINITE INERTIA FOREVER

1.   I saw a ball rolling down a hill, and as it did it gained in inertia.

2.   My hypothesis is that if there is an object in outer space, and it had a self generating motor that reacted to acceleration by increasing acceleration, then it would get faster infinitely.

3.   By using a rock, I will show how it gains speed.

4.   I was right, because it matter gains speed at all, then there is no limit.

5.   More research could prove to be useful.

*

$$Object = O$$

$$Space = S$$

$$Inertia = I$$

$$O + S = I\infty$$

•

105

THERE ARE LIFE FORMS THAT HAVE LIVED FOR THOUSANDS OF YEARS

UNDETECTED.

1.   I've read theories, and interacted with my environment.

2.   What I hypothesize is that in other dimensions there is life that could potentially be alive, and for a very long duration of time. Do increasing dimensions mean increasing life?

3.   As far as dimensions go, they are not easily detected and are still theories.

4.   I was correct, that there have been theories on other dimensions.

5.   Advanced technology could uncover more.

*

Other dimensions = O

Life = L

$$\frac{O = L}{2}$$

•

106

ADVANCES IN CAR BLINKERS

1.   I noticed, by sight, that car blinkers have room for advancement and potential.

2.   The hypothesis that I to present is one of importance and of accomplishment. Cars have

blinkers, but they could act as more and cars themselves could be more.

3.   What I plan on doing is make a diagram, graph and list of possible improvements on the

design of automobiles.

4.   I was right, based on historical fact that shows gradual advancement, there will be

improvements.

5.   More can be done, by trial and error, to meet criteria and expectations.

Car Blinkers = B

Additional Functions = A

Advancement = D

Electrical Current = C

$$B + A + C = D$$

•

SNOWBALL ROLLING EFFECT

1.   When matter rolls against matter, it gains in mass in subatomic particles.

2.   What I hypothesize is that everything in life can be in a condition of velocity and roll, under the right circumstances.

3.   I plan on rolling snow down a hill, to prove that matter acts in that way in all circumstances.

4.   I was correct, it gained mass as it rolled, becoming heavier.

5.   Everything could be studied, as to how this might benefit the human race.

•

Original Mass = O

Velocity = V

Increase In Original Mass = I

$$O + V = I$$

108

MORAL REASONING BY EXAMPLE

1.   I have noticed patterns in behaviors.

2.   My hypothesis is that people do things for different reasons, an example is when people kill fish ignoring the immoral behavior and are reaffirmed with praise from people that were not killed that day.

3.   I plan on killing an insect, then receiving praise.

4.   I was right, by committing an immoral act and then receiving praise, it ceases to be immoral and was based on opinion.

5.   No more tests are to be conducted.

Repetitive Wrongs = R

Praise = P

More Wrongs Or Rights = M

$$R + P = M$$

•

FORESTS AND PATTERNS

1. From other people's notes, I have learned of fractals and have seen first hand patterns in leaves and wood. Wood has rings and leaves branching.

2. My hypothesis is that these patterns could be lead us to a deeper understanding of why nature is how it is, and what medical and structural benefits it has to offer.

3. By taking samples off the forest floor, I will look at them under a microscope and make comparisons.

4. I was right, there were more patterns to be found and I left it alone.

5. There will be no more studies, because nature should be left alone.

Samples Off Floor = S

Patterns = P

$$S = P$$

•

GOLDEN RATIOS AND CHANNELING MATTER

1.   Though it has been rare, I have seen golden ratios and also sketched them with a pencil.

2.   My hypothesis is that as a ratio, and spiral, energy could be channeled into a machine and then increase in heat, creating an atomically related explosion that could possibly tear a hole in time and space.

3.   What I plan on doing here is condensing energy, and make it break through to another dimensional place by use of a machine.

4.   It was a success, because what I saw was suggestive of a success.

5.   Another machine, one built at another scale and perfection will provide more extensive results.

Light = L

Resulting Data = RD

$$2L * 3.14 = RD$$

•

111

MAGNETIC SPACE MACHINE, FUEL-LESS

1.   Magnetism is longer lasting and less disruptive.

2.   My hypothesis is that a space craft could be created in a more efficient way.

3.   By creating models in a vacuum, I will show the potential.

4.   It worked, with no one inside.

5.   A larger model should be invented, and more tests could proceed.

*

Magnetic Machine = M

Distance Of Machine = D

Velocity = V

$D\infty + V\infty = M$

•

112

LINK UP

1.   I have seen light move.

2.   Matter, going in exactly different directions, will combine with other matter in a
sequential method.

3.   The whole of everything could be like a mirror and wall, and there could be connections
going in or out.

4.   It was true that some matter and light connect, in an opposite and similar way.

5.   There are probably parts of the universe that have it happen on a larger scale, and it
should be observed it chance be.

*

Separate 1 = S

Separate 2 = E

Combinations = C

$$S + E = C$$

•

113

TIME CAN BE MEASURED BUT NOT CONTAINED

1.   Matter has been seen in motion, but it is seemingly out of the bin and not contained.

2.   My hypothesis is that matter will never be contained, or it is contained and the container is not visible.

3.   By containing food inside a container, I will demonstrate that if matter could be contained on a smaller scale, then it could on a much larger and much smaller scale.

4.   Demonstration completed, it was true that it happened at all in this universe, even if it was on a small scale.

5.   Through time, more will be discovered, which is my prediction.

*

Matter = M

Not Contained = N

M = N

●

114

WHEN A CHEMICAL REACTION HAPPENS IN ONE AREA OF THE UNIVERSE, IT
HAPPENS AT THE SAME EXACT TIME SOMEWHERE ELSE

1.   I've seen chemical reactions, and everything is the like.

2.   What I hypothesize is that when a chemical reaction happens in France, and also in the
U.S. it is not by chance at all.

3.   By using an alarm in one room of my house, and then another in another room, I hope to
prove that the two are very related.

4.   It worked, by using alarms I showed that two separate chemical reactions could happen at

the same time, related.

5.   More proof could be obtained.

*

Chemical reactions = C

Non-Visible Connections = I

C = I

•

115

IN CAUSALITY, WHEN TWO THINGS COLLIDE IN OUTER SPACE MORE

DIMENSIONS ARE CREATED INFINITELY

1.   Echos have been heard, and when I heard the echos and seen the waves of the ocean, I

knew there was something to it.

2.   My hypothesis is that when there is a collision of anything, that there is infinite aftermath

that lasts forever.

3.   By throwing a rock in the water, I watched the effect.

4.   I was right, that matter that collides with other matter causes a lasting effect, possibly

infinitely, because it has been proven that energy can not be destroyed.

5.   No more tests need to be conducted, at this time.

$$\text{Causing Collision Of Water} = C$$

$$\text{Effect} = E$$

$$C = E\infty$$

•

116

ONE STAR OR PLANET COULD BE BLOCKING ANOTHER, AND SO THERE COULD

BE MORE PLANETS NEARBY THAN IS APPARENT

1.   By observation, there are blind spots in human vision.

2.   A hypothesis, or mine rather, is that there are places that can not be seen, and could

provide new discovery.

3.   I plan on conducting an experiment, to show that there are blind spots on Earth.

4.   It was true, that by blockage, what can not be seen, is not known.

5.   Galactic blind spots will show us, as a human race, more.

Places Unseen By Earth Inhabitants = P

Reality = R

$$P = R$$

•

117

OTHER OBJECTS COULD BE BEHIND OTHER OBJECTS AND HIDDEN

1.   Under the surface of the Earth, in outer space and on other planets and suns, there is

uncharted territory.

2.   My hypothesis is that there could only be 10% explored, leaving 90% unexplored.

3.   What I plan on doing is gathering data and estimating that approximate remainder.

4.   More or less, it was true that there is always going to be uncharted territory.

5.   More tests could be conclusive.

Obstructions = O

View = V

Limited Sight = L

$$O + V = L$$

●

118

CHRISTIAN MUSIC CREATES HAPPIER PEOPLE

1.   Music has an influence on me, and if it is calming then I am calm.

2.   The hypothesis is that by listening to the correct music, there are a greater degree of

benefits.

3.   By listening to calming music, I will prove that it has an influence.

4.   I was correct, it brought a peacefulness, and that was how I wanted to be.

5.   There will be more studies.

●

Peaceful Music Harmonies = P

Discordant Music = D

Optimal Lifestyle = O

$$P - D = O$$

119

OUR DNA ORIGINATED ELSEWHERE IN THE UNIVERSE

1.   Our planet is like a gigantic meteor, and it came from somewhere.

2.   My hypothesis is that planet Earth came from another planet that had life at one point.

3.   By taking a piece of bread, and dividing, I will prove that Earth is like this, separating and dividing with ingredients the same.

4.   Proven by example, our planet is like the bread, it has the same ingredients.

5.   Many more studies will conclude, and support my belief.

$$DNA = D$$

$$Uncharted\ Territory = U$$

$$Chemical\ Reaction = C$$

$$D + U = C$$

•

120

ARE MORALS AND IDEAS BIOLOGICAL?

1.   By reading about morals, and having ideas they seem to be biological.

2.   All thought is biological, and originates naturally. This is my hypothesis.

3.   By taking samples of human tissue, I will show that all thought is in direct correlation

with biology.

4.   The samples proved that biology was connected with thought, and a soul.

5.   More studies will give more insight on the subject.

Thought = T

Biological = B

T = B

●

121

A CERTAIN AMOUNT, LARGELY, COULD CAUSE A REACTION NEVER BEFORE

SEEN

1.   I have seen large pieces of rock in outer space.

2.   Greater pieces of rock, or material, could collide causing very large explosions that could be a thousand times that of an atomic bomb.

3.   By using small fire crackers, I will show an explosion suggesting larger ones that have never been seen.

4.   I was right, to think that explosions happen at all scales, because it only makes sense that way.

5.   A lot more research should be given to the subject.

Chemical Reaction = C

Impact = I

Mass Explosion = E

$$C + I = E$$

•

122

MASS TYPE ONE CAN TURN INTO MASS TYPE TWO BY MEANS OF A CHEMICAL REACTION

1.   Shapes make up our universe and reality, and this has been seen, felt and heard. One

mass can alter to another, suggestively.

2.   What I hypothesize is that Mass 1 can turn into Mass 2 by means of a chemical reaction,

and therefor one thing can become another.

3.   By taking water, I will freeze it, turning it into a solid ice mass.

4.   The experiment went well, water became something else, and that means that other

masses could become other things.

5.   By conducting tests, new masses never seen could be revealed.

Mass One = O

Mass Two = T

Chemical Reaction And Transition = C

O + T = C

•

MATTER CAN BE TAKEN FROM ONE DIMENSION AND PUT INTO ANOTHER DIMENSION

1.   Through reading about them, and living among them, it seems that dimensions intersect.

2.   While in the fourth dimension, the matter in the fifth or sixth dimension could intersect, and interact.

3.   What I will do, is put my hand in water, and watch it go from being surrounded by Oxygen, to being surrounded by Hydrogen and Oxygen. Dimensions are the same way.

4.   It was true, that there can be intersections, relatively.

5.   Evidence of new findings will no doubt be uncovered.

4th Dimension = D

5th Dimension = M

6th Dimension = S

Interaction = I

$$D + M + S = I$$

●

GLOBAL WARMING AND ICE AGES ARE BOTH PREVENTABLE

1.   There have been observations, of balance and extremes in temperature.

2.   What I hypothesize is that global warming and abundances of ice could both solve the

other, problematically.

3.   By using ice, and a match, I will prove that this is true.

4.   It is true, the flame melted the abundance of ice, and the ice put out the flame.

5.   More could be conclusive.

Ice = I

Heat = H

Balances = B

$I + H = B$

•

OBJECTS NEVER TOUCH, NOTHING TOUCHES ANYTHING

1.　　As what has been seen and demonstrated, two objects either do not touch or do not bond -
and even when they bond do not touch.

2.　　A hypothesis is that on any scale, large or small, objects do not touch whether it is matter
on a massive or micro scale. The energy around the objects could bring about friction, but not the
matter itself.

3.　　By using magnets, I'll show that two things could be effected without touching.

4.　　My tests were conclusive, and showed that without touching, they could appear to be.

5.　　Use of advanced microscope technology could prove further that nothing touches
anything.

Object One = O

Object Two = T

Impact = I

$$O + T \neq I$$

•

126

A PERSON'S APPEARANCE CAN CHANGE EVERYTHING PART II

1. I've noticed that a person's appearance can effect a persons mood, social status, chances of success and financial winnings.

2. I hypothesize that by wearing certain attire, a persons life could be drastically altered, whether in a positive or negative way.

3. As I walked in the city, I wore what was commonly thought as classy clothing, and asses other peoples reactions, my reactions, confidence and manage of finance.

4. What I had thought was correct, that everything can be subject to appearance. Others were more drawn to me, I felt confident and I managed my money with more caution.

5. Further tests could conclude more.

Wearing Certain Attire = W

Benefit = B

W = B

•

127

THERE IS AN OTHER DIMENSIONAL BOWL THAT HOLDS MATTER SEEMINGLY

SUSPENDED

1.   As I went about my day, I saw and felt a fruit bowl, and noticed that the shape could be found in nature, like a half of a coconut for instance.

2.   Matter could be directed and controlled by gravitational force, and so a bowl, one that could be any size could exist in time and space. This, is my hypothesis.

3.   By using a bowl, I put matter inside it, simulating that it was like that of the shape gravity could produce.

4.   As thought previously, the bowl shape that occurred in nature was suggestive, and could also appear elsewhere in the universe.

5.   By using other objects, I plan on creating simulations to represent what could be, is, and

always would be.

Gravitational Bowl = G

Outer Space = O

Contents = C

Product = P

$$(G + B)(O) = P$$

•

128

EXTREME HEAT COULD THEN BE QUICKLY COOLED TO CREATE ANOTHER

PRODUCT

1.   When I have seen matter heated, it could be shaped if a metal and then cooled quickly.
Doing so creates an end product that takes shape.

2.   What I think, is that in the process of cooling, new atomic structuring could result.

3.   By heating water, and then quickly cooling it, I'll prove that there is an interesting effect.

4.   The end product was conclusive, because the bowl that was used cracked suggesting that there was expansion.

5.   By using other materials, more knowledge could be obtained.

$$\text{Lowering Of Temperature} = L$$
$$\text{Heightening Of Temperature} = H$$
$$\text{Result} = R$$

$$L + H = R$$

•

129

TRANSPORT ON ROPE OR TELEPORT

1.   As I sat in a chair in the city I saw the cars, buses, and other modes of transport.

2.   My hypothesis is that new transport could alter the lives of everyone in the world, and I think that it could be done by using gravity, magnetism and dimensional transitioning.

3.   By taking what is already known by man, I will construct a model that could change the

lives of everyone alive.

4.   It succeeded, using a brand new construct, travel is now a step closer to being my

efficient and also cost effective.

5.   New designs could prove as constructive.

Magnetism =M

Transport = T

M = T

•

130

CAUSALITY EXPLAINS EARTH AND EVERYTHING ON IT INCLUDING THE MILKY

WAY

1.   Cause and effect, shows that the whole galaxy is like a series of dominos in effect, and

that matter and energy are in a constant state or chemical reaction.

2.   What I hypothesize is that all of reality that we know it, is simply a series of chemical reactions, and that one object of matter simply reacts to another.

3.   Taking a ball, I will show that by throwing it in the dirt, it creates indentations. This will prove that there is only cause and effect.

4.   It was true, that like a domino effect, the ball effected the dirt, and therefor represented all other matter.

5.   More tests will show more of the same.

Giant Piece Of Matter = G

Giant Reactions = R

Real = E

$$G + R = E$$

•

131

NEUROLOGICAL CURES

1. When a chemical disturbance is seen in the brain, it is apparent.

2. My hypothesis is that if the brain has a chemical action, then another chemical could reverse it with a reaction.

3. By studying chemistry and brain activity, I will prove that there are chemicals that can create a reaction.

4. I was correct, that certain medicines cause a reaction, and suggest that there is a cure that has not yet been found for everything.

5. Many more studies will show much more about the subject.

Chemical With Negative Effects = N

Chemical With Positive Effects = P

Chemistry = C

$$N + P = C$$

•

IN MEDICINE THERE ARE AILMENTS AND CURES, ONE NOT WITHOUT THE OTHER USUALLY

1.   I have seen cancer patients, and patients will other ailments.

2.   I hypothesize that for every ailment there is a cure, because for every chemical action there is a reaction.

3.   By using chemicals, I will attempt to cure cancer by knowing that there must be a cure if there is an ailment to begin with.

4.   I was successful in attempting to find a cure.

5.   More attempts will uncover cures to ailments.

•

Ailment = A

Cure = C

Correction = O

$$A + C = O$$

$$133$$

ANYTHING THAT REPEATS DOES SO DIFFERENTLY, IN THE SLIGHTEST OR MOST,

EVERY TIME.

1.   By observing, listening and feeling, I've noticed that everything is new because it is.

2.   Nothing repeats, because everything is new. That which repeats could be only similar but

not exactly the same.

3.   Through taking an example of sound waves, I will study the frequency of the first, and

second. This is an example of sound.

4.   The first waves were slightly different than the next, and so both had novelty.

5.   More will conclude.

$$Repetition = R$$

$$Reality = E$$

$$R\infty = E$$

•

PATTERNS IN SIGHT SHOW THE POSSIBILITY OF A SYSTEM

1.   Observationally, there have been patterns that show up in a system, such as the universe.

2.   What I hypothesize is that the patterns could be studied, and more data could be retrieved.

3.   By documenting what I see and hear, and creating with it information, I will find revealing patterns.

4.   It was true, that no matter how unordered, there was order. Patterns emerged that suggested that there were only a series of explosions that shaped the cosmos.

5.   In disorder, order can be found and the universe has it. Patterns could explain more if studied than if not.

Patterns = P

Order = O

P = O

•

135

EVERYTHING IS MADE UP OF SHAPES OF THE SAME SUBSTANCE, FREQUENCIES
AND VIBRATION.

1.   When looking at pieces of scrap wood, I could see all the sizes. Matter, on a medium

scale, took form.

2.   My hypothesis is that on a microscopic scale there are shapes of matter that make up the

universe, the Earth, and people.

3.   By using pieces of wood, I will prove that if on a larger scale, it must also be so on a

smaller one.

4.   I was successful in using larger models to represent smaller ones.

5.   There will, no doubt, be more insight into this subject.

Matter On A Microscopic Scale = M

Larger Objective Matter = L

Total Matter = T

$$M + L = T$$

•

136

THAT WHICH IS TAUGHT IS LIMITING, AND PREVENTS FURTHER BREAKTHROUGHS

1.   By reading, and listening to teachers, I noticed that knowledge has limitations.

2.   What I think, and what my hypothesis is, is that knowledge is limiting and that sections of mathematics and science in particular are limiting.

3.   By using my skills, I plan on attempting to expand on mathematics and science by use of a pen.

4.   I was successful in attempting to do so. Knowing that it is possible, is proof in itself.

5.   Mathematicians could expand on techniques that have already been instated.

Study = S

Old Material = O

Fresh Outlook = F

S - O = F

●

137

THE NOVELTY

1.   Experiencing life, it has many qualities.

2.   Life experienced for the first time, provides novelty in every area of existence. My hypothesis is that there will be more times like this after death.

3.   By watching an insect die, I will study it with equipment.

4. If life is motion, then that which was dead was also a part of that motion and so my hypothesis was correct.

5. More to come, which will be insightful.

$$\text{Life} = L$$

$$\text{Death} = D$$

$$\text{Energy} = E$$

$$L + D = E$$

•

138

FRESH LOOK

1. There is a system in science, that I have seen.

2. Rename the other terms in science by my terms because then that way others will see science with a fresh new look.

3. By using different terms, in a different language, I hope to have a different outcome.

4.   I was right, that by redefining and re-labeling, it was looked at with a fresh look.

5.   Others could do this, and it would take a matter of hours.

Old Terms = O

New Terminology = N

Brand New Look = B

$$O + N = B$$

•

139

SPACE THAT CAN BE SEEN, OUTER SPACE, WILL ONE DAY BE FILLED WITH
MATTER THAT WAS ONCE CONDENSED - EXPANDED

1.   I've seen solid matter, and have seen empty space.

2.   My hypothesis is that one day all space will be taken up with matter, and that it was at
one time filled completely with matter that is now condensed.

3.   By taking a sponge as an example of matter, i'll put it in a container. Then will water, I will apply to the sponge, and this will hopefully prove my point.

4.   It was true, that the sponge filled the whole container and so matter was the same.

5.   In time, it is predicted that more information will be accessible.

Mass = M

Sponge Reaction = S

Zero Space And All Mass = Z

M + S = Z

•

140

OTHER PEOPLE ARE JUST PARTS OF YOUR SELF

1.   I saw numerous people, and they were talking and walking.

2.   What I hypothesize, is that the people we see are only parts of the self and therefor there is only one person alive in the universe witnessing a falsified dream state.

3. By observation and taking notes, I plan on proving that this is correct.

4. It was a truth! People can only be seen if there is a beholder, and without one there is no sight of anyone.

5. Waking life and dreaming life could very well continue after one dies, and more research will take place while I live and after I am not alive.

Other People = O

Self = S

Background Inanimate Objects = B

Total Experience = T

$$O + S + B = T$$

•

141

DYING IS A CHEMICAL PROCESS, AND WHEN DEAD CELLS DIVIDE AND OUR EXPERIENCE IS ON A SMALLER SCALE, WHICH INTERACTS WITH LIGHT AND SOUND

1.   Watching a plant decompose, it falls apart into sections.

2.   What I hypothesize is that the energy in the plant, leaving the plant, divides into sections. It is thought that humans are the same.

3.   By watching a plant decompose, I will determine if it does in fact divide, by means of matter and of energy as well.

4.   It was true, from the information that could be obtained. Life, as a whole, divides into energy and into sub groups and reformulates.

5.   More will come, and time will tell more.

Human Subject

Divided Energy

$$\frac{H + D}{2}$$

•

HUMAN BRAIN TRANSPLANTS

1. It has been studied before I had thought about it. Seeing brains, and the anatomical structure, is very suggestive.

2. Hypothetically, if an animal could have brain transplants, then humans could too.

3. By studying a list of subjects, and data, I will prove that human brain transplants are possible.

4. It was true, because animals have anatomically similar structures therefor showing the potential of humans with the procedure, and so I was correct.

5. Conducted studies could be of use, further more.

Animals = A

Humans = H

Brain Plant Possibility = B

$A = B$

$H = B$

$A \neq H$

•

143

HAVING SEX WITH SOMEONE THAT IS NOT KNOWN WELL LEADS TO MORE VIOLENT BEHAVIOR

1. A subject that I knew, had sex with someone they did not know well. He displayed anger afterwards and violent tendencies.

2. My hypothesis is that if sex is premature, and too soon, then the subjects could behave more erratically.

3. By studying a subject, Subject A, I will determine if he is more animal like.

4. It's true, that having sex with someone that you do not know leads to violent behavior.

5. It is usually the case, that violence can be avoided with discretion.

Unplanned Sexual Encounter = U

Possible Violence = P

$$U = P$$

•

144

GREY MATTER, YELLOW MATTER, AND PURPLE MATTER

1.   Matter has been seen emitting different frequencies, known as colors or constructs.

2.   I hypothesize that there are different kinds of matter that will someday be discovered.

3.   By making predictions, I will determine if I can.

4.   I made predictions, successfully.

5.   If something is thought of, it can lead to prediction.

Grey Matter = G

Yellow Matter = Y

Purple Matter = P

All Matter = A

$$\frac{G + Y + P = A}{2}$$

•

145

BY SAYING WORDS, THEY ECCHO IN THE MIND

1.   Hearing what is said to me, it sounded like 'Trip'.

2.   What I hypothesize is that by hearing the word Trip, I will be likely to repeat the word, and will especially if it is vocalized.

3.   By using a Subject, I will test to see if it is true.

4.   It is true.

5.   Evidence is suggestive.

Sound Waves = S

Echoing = E

$$S = E$$

•

146

WIRL POOLS HAPPEN IN THE WATER, AIR, AND EVERYWHERE ELSE. IF IT OCCURS IN ONE PLACE THERE IS MORE OF IT ELSEWHERE.

1. I have seen spirals.

2. By seeing Spirals in everything, I think they are reoccurring.

3. Observations will reveal what is or is not reoccurring.

4. It was true, evidence shows that they are reoccurring. Other spirals could also coexist.

5. More spirals could be evidential.

$$Spirals = S$$

$$S = S$$

•

147

PEOPLE SHOULD ALL BE TREATED LIKE ACTORS, AND PAID FOR SUCH. THERE
ARE HOMELESS PEOPLE, AND ROCKET SCIENTISTS THAT SHOULD BOTH BE PAID.

1.   I've seen people on stage, and off stage.

2.   My hypothesis is that people should be paid for being alive, and forced into social

situations.

3.   By using a subject, I will test my theory.

4.   Positive results, acting was a behavior whether on or off stage.

5.   Pay to everyone could bring about solutions.

Actors = A

None Actors = N

People = P

$$A + N = P$$

•

148

EVOKING ANGER IN SUBJECTS 'A' CAUSES THEM TO ATTACK INNOCENT SUBJECTS 'B'

1. I watched on person aggravate another, and the other lashed out.

2. My hypothesis is that if a person is angered, then they will vent the anger on an innocent subject.

3. By using two subjects, I will make a demonstration.

4. It was very conclusive.

5. More could bring further conclusions.

Aggravation = A

Venting Anger = V

$$A = V$$

•

149

ALCOHOL SHOULD NOT BE CONSUMED IN A COURT HOUSE

1.   Judges have drank alcohol in front of me, and I noticed that she was not performing as well.

2.   My hypothesis is that drinking in a courthouse is counterproductive.

3.   By taking notes, I will find logic to prove my point.

4.   It was a truth, that people in a courthouse perform better without alcohol consumption.

5.   No more evidence is required.

Drinking = D

Courthouse = C

Lessoned Performance = L

$$D + C = L$$

•

150

CHALLENGING TRADITION IS PROGRESSIVE

1.   By testing traditions and habit, I've found new sides to the world.

2.   My hypothesis is that by pressing against the boundaries of the usual, new notions can be discovered. Hypotheses can be pressed, expanded and extended.

3.   As I tried to do so, I succeeded in having more findings.

4.   Correctly, I was an proved my hypothesis.

5.   Boundaries can be anywhere, and testing them could produce more data.

Boundaries in number = B

$$B = B$$

•

About the author

Aaron Hatcher is a writer, visual artist, and a musician. He has written many books and screenplays, including The Vegetarian Beetle, The Vegetarian Beetle Part II, The Forest Life, The Forest Life Part II, The Man Made Entirely Out Of Paint, The Man Made Entirely Out Of Paint Part II, In Fond Memory Of Benjamin Harp, In Fond Memory Of Benjamin Harp Part II, The Spearmen, The Spearmen Part II, The building On 14th and Happy, Scientific Theories, Scientific Theories Part II, and Politics. 2012-2014. He is also known for his 2012-2013 painted works which was a success. Further on, the author is the lead singer, songwriter and guitarist for the band Lightning, albums Twinkling Stars, She's So Pretty, In The City and Do It. Screen Plays 1 - 11.